ENERGY SECTOR STANDARD OF THE PEOPLE'S REPUBLIC OF CHINA

中华人民共和国能源行业标准

Code for Design of Hoists for Hydropower Projects
Part 3: Code for Design of Screw Hoists

水电工程启闭机设计规范
第3部分：螺杆式启闭机设计规范

NB/T 10341.3-2019

Replace DL/T 5167-2002, SD 297-88, SD 298-88

Chief Development Department: China Renewable Energy Engineering Institute
Approval Department: National Energy Administration of the People's Republic of China
Implementation Date: July 1, 2020

China Water & Power Press

中国水利水电出版社

Beijing 2024

All rights reserved. No part of this publication may be reproduced, stored in a retrieval system, or transmitted in any form or by any means—electronic, mechanical, photocopying, recording or otherwise, without prior written permission of the publisher.

图书在版编目（CIP）数据

水电工程启闭机设计规范. 第3部分, 螺杆式启闭机设计规范 : NB/T 10341.3-2019 = Code for Design of Hoists for Hydropower Projects Part3:Code for Design of Screw Hoists(NB/T 10341.3-2019) : 英文 / 国家能源局发布. -- 北京 : 中国水利水电出版社, 2024. 3. -- ISBN 978-7-5226-2681-9

Ⅰ. TV664-65

中国国家版本馆CIP数据核字第202408MQ34号

ENERGY SECTOR STANDARD
OF THE PEOPLE'S REPUBLIC OF CHINA
中华人民共和国能源行业标准

Code for Design of Hoists for Hydropower Projects
Part 3: Code for Design of Screw Hoists
水电工程启闭机设计规范
第3部分：螺杆式启闭机设计规范
NB/T 10341.3-2019
Replace DL/T 5167-2002, SD 297-88, SD 298-88
（英文版）

Issued by National Energy Administration of the People's Republic of China
国家能源局　发布
Translation organized by China Renewable Energy Engineering Institute
水电水利规划设计总院　组织翻译
Published by China Water & Power Press
中国水利水电出版社　出版发行
　　Tel: (+ 86 10) 68545888　68545874
　　sales@mwr.gov.cn
　　Account name: China Water & Power Press
　　Address: No.1, Yuyuantan Nanlu, Haidian District, Beijing 100038, China
　　http://www.waterpub.com.cn
中国水利水电出版社微机排版中心　排版
北京中献拓方科技发展有限公司　印刷
184mm×260mm　16开本　4印张　127千字
2024年3月第1版　2024年3月第1次印刷
Price（定价）：￥660.00

Introduction

This English version is one of China's energy sector standard series in English. Its translation was organized by China Renewable Energy Engineering Institute authorized by National Energy Administration of the People's Republic of China in compliance with relevant procedures and stipulations. This English version was issued by National Energy Administration of the People's Republic of China in Announcement [2023] No. 5, dated October 11, 2023.

This version was translated from the Chinese Standard NB/T 10341.3-2019, *Code for Design of Hoists for Hydropower Projects—Part 3: Code for Design of Screw Hoists*, published by China Water & Power Press. The copyright is reserved by National Energy Administration of the People's Republic of China. In the event of any discrepancy in the implementation, the Chinese version shall prevail.

Many thanks go to the staff from the relevant standard development organizations and those who have provided generous assistance in the translation and review process.

For further improvement of the English version, any comments and suggestions are welcome and should be addressed to:

China Renewable Energy Engineering Institute
No. 2 Beixiaojie, Liupukang, Xicheng District, Beijing 100120, China
Website: www.creei.cn

Translating organization:

Guangdong Hydropower Planning & Design Institute Corporation Limited

Translating staff:

| TAO Yundong | LIU Xilong | HUANG Jun | ZENG Gengyun |
| GONG Fang | YE Xi | CUI Yu | |

Review panel members:

LIU Xiaofen	POWERCHINA Zhongnan Engineering Corporation Limited
LI Zhongjie	POWERCHINA Northwest Engineering Corporation Limited
QI Wen	POWERCHINA Beijing Engineering Corporation Limited
LIU Qin	POWERCHINA Northwest Engineering Corporation

	Limited
LIANG Hongli	Shanghai Investigation, Design & Research Institute Corporation Limited
HU Baowen	POWERCHINA Huadong Engineering Corporation Limited
ZHANG Qingjun	China Gezhouba Group Machinery & Ship Corporation Limited
JIA Haibo	POWERCHINA Kunming Engineering Corporation Limited
LIN Zhaohui	China Renewable Energy Engineering Institute
LI Shisheng	China Renewable Energy Engineering Institute

National Energy Administration of the People's Republic of China

翻译出版说明

本译本为国家能源局委托水电水利规划设计总院按照有关程序和规定，统一组织翻译的能源行业标准英文版系列译本之一。2023年10月11日，国家能源局以2023年第5号公告予以公布。

本译本是根据中国水利水电出版社出版的《水电工程启闭机设计规范 第3部分：螺杆式启闭机设计规范》NB/T 10341.3—2019翻译的，著作权归国家能源局所有。在使用过程中，如出现异议，以中文版为准。

本译本在翻译和审核过程中，本标准编制单位及编制组有关成员给予了积极协助。

为不断提高本译本的质量，欢迎使用者提出意见和建议，并反馈给水电水利规划设计总院。

地址：北京市西城区六铺炕北小街2号
邮编：100120
网址：www.creei.cn

本译本翻译单位：广东省水利电力勘测设计研究院有限公司

本译本翻译人员：陶云冬　刘细龙　黄　峻　曾庚运
　　　　　　　　龚　芳　叶　曦　崔　玉

本译本审核人员：

刘小芬	中国电建集团中南勘测设计研究院有限公司
李仲杰	中国电建集团西北勘测设计研究院有限公司
齐　文	中国电建集团北京勘测设计研究院有限公司
柳　青	中国电建集团西北勘测设计研究院有限公司
梁洪丽	上海勘测设计研究院有限公司
胡葆文	中国电建集团华东勘测设计研究院有限公司
张庆军	中国葛洲坝集团机械船舶有限公司
贾海波	中国电建集团昆明勘测设计研究院有限公司
林朝晖	水电水利规划设计总院
李仕胜	水电水利规划设计总院

国家能源局

Announcement of National Energy Administration of the People's Republic of China [2019] No. 8

National Energy Administration of the People's Republic of China has approved and issued 152 energy sector standards including *Code for Operating and Overhauling of Excitation System of Small Hydropower Units* (Attachment 1) and the English version of 39 energy sector standards including *Code for Safe and Civilized Construction of Onshore Wind Power Projects* (Attachment 2).

Attachments: 1. Directory of Sector Standards
2. Directory of English Version of Sector Standards

National Energy Administration of the People's Republic of China

December 30, 2019

Attachment 1:

Directory of Sector Standards

Serial number	Standard No.	Title	Replaced standard No.	Adopted international standard No.	Approval date	Implementation date
...						
18	NB/T 10341.3-2019	Code for Design of Hoists for Hydropower Projects—Part 3: Code for Design of Screw Hoists	DL/T 5167-2002, SD 297-88, SD 298-88		2019-12-30	2020-07-01
...						

Foreword

According to the requirements of Document GNKJ [2015] No. 283 issued by National Energy Administration of the People's Republic of China, "Notice on Releasing the Development and Revision Plan of the Energy Sector Standards in 2015", and after extensive investigation and research, summarization of practical experience, consultation of relevant standards of China, and wide solicitation of opinions, the drafting group has prepared this code.

The main technical contents of this code include: basic requirements, design principles and requirements, loads, materials, mechanical design, structural design, electrical design, and safety.

The main technical contents revised are as follows:

This code replaces the content regarding the design of screw hoists specified in DL/T 5167-2002, *Design Specifications for Gate Hoist in Hydropower and Water Resources Projects*; and replaces SD 297-88, *Series Parameters of QL Screw Hoist* and SD 298-88, *Technical Requirements of QL Screw Hoist*.

— Adding the classification of screw hoists.

— Adding the relevant provisions on swing-type screw hoists for lifting radial gates.

— Adding the provisions on corrosion prevention.

— Adding the requirements for welded hoist frame and hoist drive box.

— Adding the design requirements for rolling-type screw hoists.

— Adding the chapter "Safety".

— Adjusting the material list of load-bearing nuts.

— Adjusting the models and series parameters of screw hoists and expanding the series of hoisting force.

— Deleting the provision "the lead angle of thread shall not be less than 4°".

National Energy Administration of the People's Republic of China is in charge of the administration of this code. China Renewable Energy Engineering Institute has proposed this code and is responsible for its routine management. Energy Sector Standardization Technical Committee on Hydropower Steel Structures and Hoists is responsible for the explanation of specific technical contents. Comments and suggestions in the implementation of this code should be addressed to:

China Renewable Energy Engineering Institute

No. 2 Beixiaojie, Liupukang, Xicheng District, Beijing 100120, China

Chief development organizations:

 China Renewable Energy Engineering Institute

 Guangdong Hydropower Planning & Design Institute

Chief drafting staff:

LIU Xilong	TAO Yundong	LIN Zhaohui	HUANG Jun
HE Gaonian	ZENG Gengyun	HE Wencai	LI Guorui
YE Baodong	LI Zhenxiang	CHEN Yewen	FU Mingwei
YE Xi			

Review panel members:

GONG Jianxin	XIAO Duanlong	CHEN Hong	YAO Changjie
FANG Hanmei	WEI Wenwei	CHEN Xia	ZHAO Fuxin
LIAO Yongping	YU Xiquan	CUI Zhi	WU Xiaoning
GAO Wei	XU Dingguang	ZHONG Xiaodong	CHEN Wenhua
LI Shisheng			

Contents

1	**General Provisions**	1
2	**Terms**	2
3	**Basic Requirements**	3
4	**Design Principles and Requirements**	4
4.1	General Requirements	4
4.2	Type Selection and Configuration	5
4.3	Transmission Mechanism	5
4.4	Hoist Frame and Embedded Parts	6
4.5	Safety Protection Devices	6
4.6	Corrosion Prevention	6
5	**Loads**	7
6	**Materials**	8
6.1	Castings	8
6.2	Forgings and Rolled Pieces	8
6.3	Structural Members	9
6.4	Connecting Materials	9
7	**Mechanical Design**	11
7.1	Lifting Screw and Load-Bearing Nut	11
7.2	Driving Mechanism and Parts	17
8	**Structural Design**	23
8.1	Calculation Principles	23
8.2	Load Combinations	23
8.3	Allowable Stresses	23
8.4	Strength Calculation for Structural Members and Connections	27
9	**Electrical Design**	33
9.1	General Requirements	33
9.2	Power Supply	33
9.3	Control and Protection Devices	33
9.4	Lighting and Signaling	33
9.5	Earthing and Lightning Protection	34
10	**Safety**	35
10.1	Markings, Nameplates and Safety Signs	35
10.2	Safety Requirements for Mechanism, Structure, Electrical and Fire Protection	35
10.3	Operation and Maintenance Manuals	36

Appendix A   Models and Series Parameters of Screw Hoists ······· 37
Appendix B   Design Profiles and Dimensions of Trapezoidal
              Screw Threads ······································· 40
Appendix C   Friction Factors and Efficiencies ················· 42
Appendix D   Stability Coefficient of Eccentric Compression ······ 43
Appendix E   Motor Power Correction Considering Working
              Environment ··· 46
Appendix F   Allowable Values of Common Sliding Bearing
              Materials ··· 48
Explanation of Wording in This Code ···························· 49
List of Quoted Standards ······································· 50

1 General Provisions

1.0.1 This code is formulated with a view to standardizing the design of screw hoists for hydropower projects.

1.0.2 This code is applicable to the design of screw hoists for operating the gates for hydropower projects.

1.0.3 In addition to this code, the design of screw hoists for hydropower projects shall comply with other current relevant standards of China.

2 Terms

2.0.1 screw hoist

equipment used for opening or closing a gate by the screwing of the lifting screw and the load-bearing nut to transmit the motion and force, transforming the rotational motion to linear motion

2.0.2 lifting screw

long rod used for transmitting loads or displacements in a screw hoist, whose stroke part is of an externally threaded cylinder

2.0.3 load-bearing nut

block supported by the hoist frame and used for bearing the load transmitted from the lifting screw, whose internal thread mates with the external thread of the lifting screw

2.0.4 lifting force

hoisting force applied to open a gate

2.0.5 closing force

hoisting force applied to close a gate

2.0.6 holding force

load borne by a hoist to control the lowering speed of a gate

2.0.7 screw drive

mechanical drive that transforms rotational motion to linear motion by screwing the lifting screw or the load-bearing nut, which is classified as sliding or rolling type according to the friction type of the thread pair

3 Basic Requirements

3.0.1 A screw hoist shall comprise a drive unit, a lifting screw, a load-bearing nut, a support structure, a detector, etc., and an electric screw hoist shall also include an electrical control system and an electrical protection system.

3.0.2 For the design of a screw hoist, the following data shall be collected:

1. Gate operation requirements for hoist operating conditions, lifting height, hoisting speed, hoisting mode, water filling mode, etc.
2. Environmental conditions, such as hydrology, meteorology, sediment, water quality, and altitude.
3. Loads.
4. Layout of hydraulic structures, dimensions of gates and slots, and relevant dimensions and requirements for connection of gates and hoists.
5. Electrical control modes and interface requirements.
6. Requirements for power source.
7. Conditions and requirements for manufacture, transportation and installation.
8. Seismic data.
9. Contract documents and other special requirements.

3.0.3 The models and series parameters of screw hoists should be in accordance with Appendix A of this code.

3.0.4 Screw hoists shall be provided with safety protection devices, including load limiter, limiting device, and electrical protection device.

3.0.5 Screw hoists shall be protected against lightning, moisture, corrosion and blowing sand, and shall be provided with ventilation, according to the operating environment.

3.0.6 The fatigue strength need not be calculated for the mechanical parts and structural members.

3.0.7 The transportation unit of a screw hoist should not exceed the maximum overall dimensions and weight specified for transportation and limited by the field installation conditions, and shall have adequate stiffness.

3.0.8 The shop assembly and detection of screw hoists shall comply with the current sector standard NB/T 35051, *Code for Manufacture Erection and Acceptance of Gate Hoists in Hydropower Projects*.

4 Design Principles and Requirements

4.1 General Requirements

4.1.1 The design of a screw hoist shall meet the requirements for safety and serviceability, operational reliability, technological advancement, economic rationality, easy maintenance, landscaping, workplace safety, energy conservation, environmental protection, structural coordination, and friendly human-machine interface.

4.1.2 The overall design and layout of a screw hoist shall be determined through a techno-economic demonstration, according to the layout of hydraulic structures, gate type, lifting height, hoisting speed, operating time, etc.

4.1.3 Screw hoists apply to small- and medium-sized gates, and each gate shall be provided with a hoist.

4.1.4 A room should be provided for screw hoists, and the requirements for the hoist room shall be specified according to the local climate, status quo of blowing sand, operation and maintenance, etc. There shall be enough space for operation and maintenance, and the maintenance access between the screw hoist and the wall should be at least 0.8 m in width. The screw hoist installed outdoors shall be equipped with a removable cover, and protection against lightning, dust, moisture and rain shall be provided for electrical equipment.

4.1.5 The stroke of the screw hoist shall meet the operation and maintenance requirements of the gate, with a certain margin.

4.1.6 The setting elevation of the screw hoist shall meet the requirements of safe operation, prevent the actuator and electrical equipment from being flooded, and allow for the maintenance of the gate, slot and hoist parts.

4.1.7 When a screw hoist is used to operate a plain gate, the centerline of the lifting screw shall be aligned with the lifting centerline of the gate. When a screw hoist is used to operate a radial gate, a flexible and reliable swing device shall be set.

4.1.8 The screw hoist with dual or more lifting points shall be provided with a synchronization device.

4.1.9 The screw hoist shall be provided with local control. The screw hoist required for remote control shall be provided with video monitoring system.

4.1.10 The electric screw hoist used for operating the flood discharging gate shall be equipped with an emergency power supply. Small- and medium-sized screw hoists should adopt both manual power and electric power. For large-

and medium-sized screw hoists, a power-free emergency manipulating device should be provided.

4.2 Type Selection and Configuration

4.2.1 Screw hoists may be classified into:

1. Manual, electric, and manual-electric, according to drive mode.

2. Vertical type, horizontal type, and inclined type, according to lifting screw arrangement.

3. Fixed type and swing type, according to the supporting structure of lifting screw.

4. Single-point type, dual-point type, and multi-point type, according to the number of lifting points.

5. Sliding type and rolling type, according to the type of screw drive pair.

4.2.2 The type selection of a screw hoist shall be determined according to the gate type, layout, hoisting force, operating requirements, etc.

4.2.3 The parameters of lifting screw and load-bearing nut shall be selected according to the transmission mode; and corresponding drive unit, supporting structure, embedded parts, electrical control system, and detector shall be configured.

4.2.4 The hoist pedestal and hoist drive box may be welded or cast, and shall meet the requirements of strength and stiffness. The drive box with lubricating oil shall be sealed.

4.2.5 The electric screw hoist shall be provided with a stroke detector, and the manual screw hoist shall be provided with a stroke indicator.

4.2.6 For the screw hoist with down force whose lifting rod is hinged, the guiding bearing seat with a friction-reduction slide bushing shall be provided.

4.3 Transmission Mechanism

4.3.1 The transmission mechanism of screw hoists should be self-locking; otherwise, brakes shall be set.

4.3.2 Screw hoists should adopt Gleason spiral bevel transmission or worm gear transmission.

4.3.3 For a manual screw hoist, the manual operation shall be easy, and the operating force applied by one person should not be greater than 150 N. The hoisting capacity of the screw hoist driven by handwheel should not be greater than 32 kN.

4.3.4 For the slide type screw drive pair, trapezoidal threads shall be employed, whose strength shall be checked.

4.3.5 The slenderness ratio of a compressed lifting screw in full extension shall not be greater than 200, and the stability shall be checked. The slenderness ratio of a tensile lifting screw in full extension shall not be greater than 250.

4.3.6 Rotating mechanical parts such as load-bearing nuts and gears shall adopt bearings according to the nature of service and operation requirements. For electric-driven rotating mechanical parts, roller bearings should be selected.

4.4 Hoist Frame and Embedded Parts

4.4.1 The hoist frame should be welded, and its design load shall not be less than the sum of 1.1 times the rated hoisting load and the dead weight of screw hoist.

4.4.2 The hoist frame shall meet the requirements for strength, stiffness and stability.

4.4.3 Embedded parts shall have enough strength and shall not fail under the rated hoisting load.

4.5 Safety Protection Devices

4.5.1 Electric and manual-electric screw hoists shall be provided with overload protection devices and limiting devices.

4.5.2 Manual-electric and manual screw hoists shall be equipped with safety handles.

4.5.3 For a manual-electric screw hoist, the manual operation mechanism and the electric operation mechanism shall be interlocked.

4.6 Corrosion Prevention

The corrosion prevention requirements shall be determined according to the working environment of the hoist, environmental protection requirements, service life, working conditions, etc. In addition, the corrosion prevention of the hoist shall also comply with the current sector standard DL/T 5358, *Technical Code for Anticorrosion of Metal Structures in Hydroelectric and Hydraulic Engineering*.

5 Loads

5.0.1 The dead load shall include the weights of structural members, mechanical equipment, and electrical equipment of the screw hoist.

5.0.2 The hoist load shall be the maximum lifting force, holding force, and closing force applied to the lifting lugs of the screw hoist connected to the gate or tie rod.

5.0.3 The wind load and snow load need not be considered in the design of screw hoist body, but shall be considered for the hoist cover, and their values shall be taken in accordance with the current national standard GB 50009, *Load Code for the Design of Building Structures*.

5.0.4 The temperature load need not be considered in the design of a screw hoist.

5.0.5 The seismic load shall be considered when the design seismic intensity is VII or above.

6 Materials

6.1 Castings

6.1.1 Carbon steel castings shall be made of ZG230-450, ZG270-500, ZG310-570, ZG340-640, or other materials stipulated in the current national standard GB/T 11352, *Carbon Steel Castings for General Engineering Purpose*.

6.1.2 Alloy steel castings shall be made of ZG35Cr1Mo, ZG42Cr1Mo, ZG40Cr1, ZG34Cr2Ni2Mo, ZG65Mn, ZG40Mn2, ZG50Mn2, or other materials stipulated in the current sector standard JB/T 6402, *Heavy Low Alloy Steel Castings—Technical Specification*.

6.1.3 Grey iron castings shall be made of HT150, HT200, HT250, or other materials stipulated in the current national standard GB/T 9439, *Grey Iron Castings*.

6.1.4 Spheroidal graphite iron castings shall be made of QT450-10, QT500-7, or other materials stipulated in the current national standard GB/T 1348, *Spheroidal Graphite Iron Castings*.

6.1.5 Malleable iron castings shall be made of KTH300-06, KTZ450-06, KTZ550-04, KTB350-04, or other materials stipulated in the current national standard GB/T 9440, *Malleable Iron Castings*.

6.1.6 Copper alloy castings shall be made of ZCuSn5Pb5Zn5, ZCuSn10P1, ZCuAl10Fe3, ZCuAl10Fe3Mn2, ZCuPb30, or other materials stipulated in the current national standard GB/T 1176, *Casting Copper and Copper Alloys*.

6.2 Forgings and Rolled Pieces

6.2.1 Carbon steel forgings shall be made of 20, 25, 35, 45, 55, 50Mn, 65Mn, or other materials stipulated in the current national standard GB/T 699, *Quality Carbon Structure Steels*.

6.2.2 Low alloy forgings and rolled pieces shall be made of Q355 stipulated in the current national standard GB/T 1591, *High Strength Low Alloy Structural Steels*.

6.2.3 Alloy steel forgings shall be made of 35SiMn, 42SiMn, 40MnB, 35CrMo, 42CrMo, 40Cr, 40CrNi, 20CrMnTi, or other materials stipulated in the current national standard GB/T 3077, *Alloy Structure Steels*.

6.2.4 Stainless steel forgings shall be made of materials stipulated in the current standards GB/T 1220, *Stainless Steel Bars*; and JB/T 6398, *Heavy Stainless Acid Resistant Steel and Heat Resistant Steel Forgings—Technical Specification*.

6.2.5 Forgings with high requirements for hardness and wear resistance shall be made of materials stipulated in the current national standards GB/T 1299, *Tool and Mould Steels*; and GB/T 18254, *High-Carbon Chromium Bearing Steel*.

6.3 Structural Members

6.3.1 Load-bearing structural members shall be made of Q235 stipulated in the current national standard GB/T 700, *Carbon Structural Steels*; or Q355 stipulated in the current national standard GB/T 1591, *High Strength Low Alloy Structural Steels*. The materials of load-bearing structural members of the screw hoist shall be selected in accordance with Table 6.3.1.

Table 6.3.1 Materials of load-bearing structural members

Operating ambient temperature	Not lower than 0℃	Not lower than -20℃	Lower than -20℃
Steel grade	Q235B, Q355B	Q235C, Q355C	Q235D, Q355D

NOTE The operating ambient temperature is determined based on the annual mean minimum daily temperature at the place where the hoist operates.

6.3.2 Stainless steel materials used for structural members shall be made of 06Cr19Ni10, 12Cr18Ni9, or other materials stipulated in the current national standard GB/T 20878, *Stainless and Heat-Resisting Steels—Designation and Chemical Composition*.

6.4 Connecting Materials

6.4.1 Welding materials shall meet the following requirements:

1. Electrodes for manual welding shall comply with the current national standards GB/T 5117, *Covered Electrodes for Manual Metal Arc Welding of Non-alloy and Fine Grain Steels*; or GB/T 5118, *Covered Electrodes for Manual Metal Arc Welding of Creep-Resisting Steels*. The type of electrodes shall be compatible with the strength of the base metal.

2. Wire electrodes and fluxes used for both automatic and semi-automatic welding shall comply with the current national standards GB/T 5293, *Solid Wire Electrodes, Tubular Cored Electrodes and Electrode/Flux Combinations for Submerged Arc Welding of Non Alloy and Fine Grain Steels*; and GB/T 12470, *Solid Wire Electrodes, Tubular Cored Electrodes and Electrode/Flux Combinations for Submerged Arc Welding of Creep-Resisting Steels*. The types of wire electrodes and fluxes shall be compatible with the strength of the base metal.

6.4.2 The materials of fasteners shall meet the following requirements:

1 General purpose bolts, screws and studs shall be made of materials stipulated in the current national standards GB/T 3098.1, *Mechanical Properties of Fasteners—Bolts, Screws and Studs*; and GB/T 3098.3, *Mechanical Properties of Fasteners—Set Screws*. Nuts shall be made of materials stipulated in the current national standard GB/T 3098.2, *Mechanical Properties of Fasteners—Nuts*.

2 Stainless steel bolts, screws and studs shall be made of materials stipulated in the current national standard GB/T 3098.6, *Mechanical Properties of Fasteners—Stainless Steel Bolts, Screws and Studs*. Stainless steel nuts shall be made of materials stipulated in the current national standard GB/T 3098.15, *Mechanical Properties of Fasteners—Stainless Steel Nuts*.

3 High-strength bolts, nuts and washers shall be made of materials stipulated in the current national standard GB/T 1231, *Specifications of High Strength Bolts with Large Hexagon Head, Large Hexagon Nuts, Plain Washers for Steel Structures*.

6.4.3 The load-bearing connecting pins should be made of Grade 35 or 45 steel stipulated in the current national standard GB/T 699, *Quality Carbon Structure Steels*; or 42CrMo or 40Cr stipulated in the current national standard GB/T 3077, *Alloy Structure Steels*. The materials should be subjected to heat treatment as necessary.

7 Mechanical Design

7.1 Lifting Screw and Load-Bearing Nut

7.1.1 For the screw hoist adopting sliding type screw drive, the lifting screw should be made of Grade 35 or 45 steel stipulated in the current national standard GB/T 699, *Quality Carbon Structure Steels*; when stainless steel is required, 12Cr13 or 20Cr13 stipulated in the current national standard GB/T 1220, *Stainless Steel Bars* may be adopted. The material of load-bearing nuts should be bronze. For the manual screw hoist with a capacity of 32 kN or below, spheroidal graphite cast iron may be used as the material of load-bearing nuts.

7.1.2 The trapezoidal threads of sliding type screw drive pair shall comply with the current national standard GB/T 5796.3, *Trapezoidal Screw Threads—Part 3: Basic Dimensions*. The lead angle shall be smaller than the equivalent friction angle, but should not be greater than 4.5°. The design profiles and dimensions of trapezoidal screw threads shall be in accordance with Appendix B of this code.

7.1.3 For the screw hoist adopting sliding type screw drive, the strength calculation of lifting screw shall meet the following requirements:

1. The friction factor and efficiency should be in accordance with Appendix C of this code.

2. The torque borne by a lifting screw shall be calculated by the following formulae:

$$M_k = F_1 \tan(\alpha + \rho') \frac{d_2}{2} \tag{7.1.3-1}$$

$$\rho' = \tan^{-1}\left[\frac{f}{\cos(\beta/2)}\right] \tag{7.1.3-2}$$

where

M_k is the torque (N · mm);

F_1 is the hoisting load (N);

α is the lead angle of thread;

ρ' is the equivalent friction angle of thread;

d_2 is the pitch diameter of thread (mm);

f is the sliding friction coefficient between the lifting screw and the load-bearing nut, which is taken as 0.07 to 0.15 depending on the material, machining precision, and lubrication, or may be taken as 0.12;

β is the profile angle of thread, taken as 0° for rectangular threads and 30° for trapezoidal threads.

3   The torsional shear stress of a lifting screw shall be calculated by the following formula:

$$\tau_k = \frac{M_k}{0.2 d_1^3} \tag{7.1.3-3}$$

where

τ_k is the torsional shear stress (N/mm^2);

M_k is the torque (N · mm);

d_1 is the minor diameter of thread (mm).

4   The bending moment borne by a lifting screw shall be calculated by the following formula:

$$M = \frac{F_1 f d_0}{2} \tag{7.1.3-4}$$

where

M is the bending moment (N · mm);

F_1 is the hoisting load (N);

f is the friction factor of swing support, which shall be in accordance with Article C.0.1 of this code;

d_0 is the hinge shaft diameter of swing support (mm).

5   The bending stress of a lifting screw shall be calculated by the following formula:

$$\sigma_w = \frac{M}{0.1 d_1^3} \tag{7.1.3-5}$$

where

σ_w is the bending stress (N/mm^2);

M is the bending moment (N · mm);

d_1 is the minor diameter of thread (mm).

6   The axial stress of a lifting screw shall be calculated by the following formula:

$$\sigma_p = \frac{4 F_1}{\pi d_1^2} \tag{7.1.3-6}$$

where

  σ_p is the axial stress (N/mm²);

  F_1 is the hoisting load (N);

  d_1 is the minor diameter of thread (mm).

7 The combined stress of a lifting screw shall be calculated by the following formulae:

$$\sigma_F = \sqrt{(\sigma_w + \sigma_p)^2 + 3\tau_k^2} \tag{7.1.3-7}$$

$$\sigma_F \leq \frac{R_s}{2.5} \tag{7.1.3-8}$$

where

  σ_F is the combined stress (N/mm²);

  σ_w is the bending stress (N/mm²);

  σ_p is the axial stress (N/mm²);

  τ_k is the shear stress (N/mm²);

  R_s is the yield strength of steel (N/mm²), which may be taken as the upper yield strength R_{eH}, the lower yield strength R_{eL}, or the offset yield strength $R_{p0.2}$.

7.1.4 For the screw hoist adopting sliding type screw drive, the stability check of a lifting screw shall meet the following requirements:

1 When the slenderness ratio of a lifting screw is greater than 100, the following formula shall be used:

$$F_2 \leq \frac{\pi^3 E d_1^4}{64n(\mu L)^2} \tag{7.1.4-1}$$

where

  F_2 is the closing force (N);

  E is the elasticity modulus of material (N/mm²);

  d_1 is the minor diameter of thread (mm);

  n is the buckling safety factor, which is not less than 2.0;

  μ is the conversion coefficient for length, which is taken as 1.0 when both ends are hinged, and 0.7 when one end is fixed but the other end is hinged;

  L is the actual length for checking the compression of lifting screw (mm).

2   When the slenderness ratio of a lifting screw is less than 100 and additional bending moment exists, the following formula shall be used:

$$F_2 \leq A\left(\frac{R_s}{2.3}\phi_p\right) \quad (7.1.4\text{-}2)$$

where

F_2 is the closing force (N);

A is the sectional area corresponding to the minor diameter of thread (mm²);

ϕ_p is the eccentric compression coefficient, which shall be determined in accordance with Appendix D of this code.

7.1.5   For the screw hoist adopting sliding type screw drive, the strength calculation of a load-bearing nut shall meet the following requirements:

1   The nut height shall meet the requirements of the bearing stress of thread which can be calculated by the following formulae:

$$q = \frac{4F_1 P}{\pi\left(d^2 - d_1^2\right)H} \quad (7.1.5\text{-}1)$$

$$q \leq [q] \quad (7.1.5\text{-}2)$$

where

q is the bearing stress (N/mm²);

F_1 is the hoisting load (N);

P is the pitch of thread (mm);

d is the major diameter of thread (mm);

d_1 is the minor diameter of thread (mm);

H is the nut height (mm); H/P is the number of duty threads, taken as 10 if H/P is greater than 10;

$[q]$ is the allowable bearing stress (N/mm²), which shall be in accordance with Table 7.1.5 of this code.

2   The bending strength of thread root shall be calculated by the following formulae:

$$\sigma_w = \frac{3hF_1}{\pi d Z b^2} \quad (7.1.5\text{-}3)$$

$$\sigma_w \leq [\sigma_w] \quad (7.1.5\text{-}4)$$

where

σ_w is the bending stress (N/mm²);

    h is the thread height (mm), taken as $0.5P$;

    F_1 is the hoisting load (N);

    d is the major diameter of thread (mm);

    Z is the number of duty threads, taken as 10 when Z is greater than 10;

    b is the thickness of thread root (mm), taken as $0.61P$ for trapezoidal threads;

    $[\sigma_w]$ is the allowable bending stress (N/mm^2), which shall be in accordance with Table 7.1.5 of this code.

3 The shear strength of thread root shall be calculated by the following formulae:

$$\tau = \frac{F_1}{\pi d b Z} \quad (7.1.5\text{-}5)$$

$$\tau \leq [\tau] \quad (7.1.5\text{-}6)$$

where

    τ is the shear stress (N/mm^2);

    F_1 is the hoisting load (N);

    d is the major diameter of thread (mm);

    b is the thickness of thread root (mm);

    Z is the number of duty threads, taken as 10 when Z is greater than 10;

    $[\tau]$ is the allowable shear stress (N/mm^2), which shall be in accordance with Table 7.1.5 of this code.

Table 7.1.5 Allowable stresses of commonly used materials of load-bearing nuts

Material	Allowable bearing stress $[q]$ (N/mm^2)	Allowable bending stress $[\sigma_w]$ (N/mm^2)	Allowable shear stress $[\tau]$ (N/mm^2)
ZCuSn5Pb5Zn5	6 - 8	30 - 40	25 - 30
ZCuSn10P1	10 - 13	40 - 55	30 - 41
ZCuAl10Fe3	15 - 20	66 - 78	50 - 58
QT450-10	10 - 12	80 - 85	50

NOTE   The minimum value shall be taken for sand casting, and the maximum value for metal die casting.

7.1.6 For the screw hoist adopting sliding type screw drive pair, the transmission efficiency shall be calculated by the following formula:

$$\eta_1 = \eta_2 \frac{\tan \alpha}{\tan(\alpha \pm \rho')} \tag{7.1.6}$$

where

- η_1 is the transmission efficiency of sliding type screw drive pair;
- η_2 is the efficiency of bearing, which shall be determined in accordance with Article C.0.2 of this code;
- α is the lead angle of thread;
- ρ' is the equivalent friction angle of thread.

NOTE   When the direction of axial load is opposite to the direction of motion, positive sign is taken; when the direction of axial load is the same as the direction of motion, negative sign is taken.

7.1.7 For the screw hoist adopting rolling type screw drive, the lifting screw should be made of Grade 55, 50Mn, or 65Mn steel stipulated in the current national standard GB/T 699, *Quality Carbon Structure Steels*; when stainless steel is required, 95Cr18 stipulated in the current national standard GB/T 1220, *Stainless Steel Bars* may be used. The material of load-bearing nut may be GCr15 stipulated in the current national standard GB/T 18254, *High-Carbon Chromium Bearing Steel*; CrWMn stipulated in the current national standard GB/T 1299, *Tool and Mould Steels*; 95Cr18 stipulated in the current national standard GB/T 1220, *Stainless Steel Bars*; 20CrMnTi stipulated in the current national standard GB/T 3077 *Alloy Structure Steels*, etc. The material of rolling element may be GCr15 stipulated in the current national standard GB/T 18254, *High-Carbon Chromium Bearing Steel*; 95Cr18 stipulated in the current national standard GB/T 1220, *Stainless Steel Bars*, etc.

7.1.8 For the screw hoist adopting rolling type screw drive pair, a shoulder, axle shaft block, etc. shall be set at the end of raceway to prevent the nut from jumping off.

7.1.9 The internal thread accuracy of trapezoidal thread shall not be inferior to Grade 9H stipulated in the current national standard GB/T 5796.4, *Trapezoidal Screw Threads—Part 4: Tolerances*; and the external thread accuracy of trapezoidal thread shall not be inferior to Grade 9c stipulated in the current national standard GB/T 5796.4, *Trapezoidal Screw Threads—Part 4: Tolerances*. The working face of thread shall be smooth, clean, and free of burrs. The surface roughness shall not be inferior to Ra6.3 μm.

7.1.10 The straightness of a lifting screw shall not be greater than 0.6 mm within 1 m. The total straightness shall not be greater than 1.5 mm for a lifting screw not longer than 5 m, and 2 mm for a lifting screw not longer than 8 m.

7.2 Driving Mechanism and Parts

7.2.1 The selection of motors shall meet the following requirements:

1 The static power of mechanism shall be calculated by the following formula:

$$N = \frac{Fv}{1000\eta} \tag{7.2.1}$$

where

N　is the static power of mechanism (kW);

F　is the rated lifting force (N);

v　is the lifting speed of gate (m/s);

η　is the total transmission efficiency.

2 The motor shall be selected according to the static power of mechanism, operating mode, and loading duration, and need not be subjected to overload and thermal check. The motors with short-time or intermittent duty for crane and metallurgical applications should be selected.

3 The degrees of protection shall meet the following requirements:

　1) The degree of protection of motor enclosure shall comply with the current national standard GB/T 4942.1, *Degrees of Protection Provided by the Integral Design of Rotating Electrical Machined (IP Code)—Classification*.

　2) The degree of protection shall be IP44 or higher for indoor application.

　3) The degree of protection shall be IP54 or higher for outdoor application. Where condensate is likely to present, the outlet of condensate shall be unblocked.

　4) When the motor is protected by external means, a lower degree of protection may be adopted.

4 When the motor operates at an altitude above 1000 m or the operating ambient temperature is inconsistent with its rated ambient temperature, the power correction of the motor shall be in accordance with Appendix E of this code.

7.2.2 Shaft couplings shall meet the following requirements:

1. The shaft coupling may be selected in accordance with the current standards of China GB/T 4323, *Pin Coupling with Elastic Sleeve*; or JB/T 8854.2, *Curved Tooth Coupling GIICL, GIICLZ* depending on operating conditions.

2. The shaft coupling shall be selected from the table of standard specification of shaft coupling according to the torque it transmits, the journal size, and the speed of the connected shaft.

3. The allowable torque of shaft coupling shall satisfy the following formulae:

$$M_L = KM_L' \tag{7.2.2-1}$$

$$M_L \leq [M_L] \tag{7.2.2-2}$$

where

M_L is the calculated torque of shaft coupling (N·m);

K is the safety factor, taken as 1.8;

M_L' is the torque transmitted from the shafts connected by the shaft coupling calculated by the rated torque of motor;

$[M_L]$ is the allowable torque of shaft coupling.

7.2.3 The brake shall be a holding brake, and the braking safety factor shall meet the following requirements:

1. For the screw hoist with only one drive, one brake shall be provided, and the braking safety factor shall not be less than 1.75.

2. For the screw hoist with two drives in rigid connection, each drive shall be provided with one brake, and the braking safety factor of each brake calculated by the total braking torque shall not be less than 1.25.

7.2.4 The speed reducer shall be of the standardized type, and the transmission ratio shall be determined by the overall transmission scheme. The speed reducer shall be selected by the rated load or the rated motor power and the required operating conditions, and the maximum radial load at the output shaft end of the speed reducer shall be checked if necessary.

7.2.5 Bearings shall meet the following requirements:

1. For sliding bearings, the maximum unit pressure p, as well as the product pv, where v is the relative linear velocity of the rotating faying surface, shall be calculated. Meanwhile, p, v, and pv shall not exceed

their respective allowable values. The allowable values of common sliding bearing materials shall be in accordance with Appendix F of this code. The sliding bearings should be of self-lubricating type.

2   For rolling bearings, the equivalent dynamic or static load shall be calculated according to the following conditions, to determine the rated dynamic or static load and the rolling bearing model:

1) For rolling bearings with a speed less than 10 r/min, only the rated static load is calculated.

2) Imposed radial load.

3) Imposed axial load.

4) Operating conditions, nature of loads, rotating race, type and supply method of lubricating oil.

5) Structural type and outline dimensions of bearings.

7.2.6 Load limiters shall meet the following requirements:

1   The combined error shall not exceed 5 %. An audible and visual alarm signal shall be given when the hoisting load reaches 95 % of the rated load of the hoist. During gate hoisting, when the hoisting load is between 100 % and 110 % of the rated load, an overload alarm signal shall be given and the power supply shall be automatically cut off.

2   For the hoist with multiple lifting points, each lifting point shall be provided with a load limiter.

3   The calibration value of lifting force and closing force of a gate shall be adjusted according to the hoisting needs of the gate, and the lifting force and closing force may be limited separately if necessary.

7.2.7 Stroke detectors and limiting devices shall meet the following requirements:

1   The stroke detector shall have a good anti-interference performance, and the output signal of sensor shall be continuous, reliable, and good in anti-electromagnetic interference.

2   The stroke detector should adopt a sensor with absolute output signal. When a sensor with incremental output signal is used, the electrical control system shall be provided with an uninterrupted power supply with a sufficient capacity according to the operation requirements of the hoist.

3   Position control: the limiting device shall automatically cut off power and give a signal when the upper or lower limit is reached.

7.2.8 The design of mechanical parts shall meet the following requirements:

1   The allowable stresses of mechanical parts shall be in accordance with Table 7.2.8.

2   Gears shall meet the following requirements:

1) The gear drive of transmission mechanism shall be of the enclosed type, and the bevel gear should be Gleason spiral bevel gear with a hard tooth surface.

2) The materials of gear shall be in accordance with the current national standards GB/T 699, *Quality Carbon Structure Steels*; and GB/T 3077, *Alloy Structure Steels*. The material selection and hardness after heat treatment shall meet the requirements of gear matching. In the case of soft or moderately hard tooth-surface gear pairs, the tooth surface hardness of pinion shall be greater than that of the main gear, and the tooth surface hardness difference shall be taken as 30 HBW to 50 HBW.

3) The tooth surface contact strength and gear bending strength shall be calculated for gear drives.

3   Worms and wormwheels shall meet the following requirements:

1) The strength calculation of worm and wormwheel shall be based on the contact strength calculation of gear tooth surface of the wormwheel, and the bending strength of gear tooth surface shall be calculated for check. When the worm doubles as a transmission shaft, it shall be taken as a shaft for strength and stiffness calculation.

2) The wormwheel should be made of copper-based alloy or zinc-based alloy. For a small screw hoist, the wormwheel may be made of spheroidal graphite cast iron. The material of worm may be quality carbon structural steel or structural alloy steel.

4   Shafts shall meet the following requirements:

1) Shafts should be made of Grade 35 or 45 steel stipulated in the current national standard GB/T 699, *Quality Carbon Structure Steels*; or 35CrMo, 42CrMo, 40Cr, 40CrNi, 35SiMn, 42SiMn, or 40MnB stipulated in the current national standard GB/T 3077, *Alloy Structure Steels*.

Table 7.2.8 Allowable stresses of mechanical parts (N/mm²)

Stress type	Symbol	Carbon structural steel	Low alloy steel	Quality carbon structural steel		Cast carbon steel					Alloy cast steel			Structural alloy steel	
		Q235	Q355	35	45	ZG230-450	ZG270-500	ZG310-570	ZG340-640	ZG50Mn2	ZG35Cr1Mo	ZG34Cr2Ni2Mo		42CrMo	40Cr
Tensile, compressive or bending stress	$[\sigma]$	100	145	135	155	100	115	135	145	195	170 (215)	(295)		(365)	(320)
Shear stress	$[\tau]$	60	85	80	90	60	70	80	85	115	100 (130)	(175)		(220)	(190)
Local compressive stress	$[\sigma_{cd}]$	150	215	200	230	150	170	200	215	290	255 (320)	(440)		(545)	(480)
Local contact compressive stress	$[\sigma_{cj}]$	80	115	105	125	80	90	105	115	155	135 (170)	(235)		(290)	(255)
Tensile stress of hole wall	$[\sigma_k]$	115	165	155	175	115	130	155	165	225	195 (245)	(340)		(420)	(365)

NOTES:
1 The value in the bracket is the allowable stress after quenching and tempering.
2 The value of $[\sigma_k]$ is corresponding to the fixed type connection, and shall be lowered by 20 % otherwise.
3 The allowable stresses of structural alloy steel are applicable to the mechanical parts with a sectional dimension of 25 mm. Considering the impact of thickness, the allowable stresses may be decreased in proportion to the decrease of yield strength.
4 The allowable stresses of cast carbon steel are applicable to the steel castings with a thickness of 100 mm or below.

2) The size of shafts shall be preliminarily determined by the allowable stress method.

3) When the speed of drive shaft is greater than 400 r/min, in addition to the strength and stiffness calculation, the critical speed shall be checked and satisfy the following formulae:

$$n_{max} \leq \frac{n_{cr}}{1.2} \qquad (7.2.8\text{-}1)$$

$$n_{cr} = 121\frac{\sqrt{d_2^2 + d_1^2}}{L^2} \qquad (7.2.8\text{-}2)$$

where

n_{max} is the actual maximum speed of shaft (r/min);

n_{cr} is the critical speed (r/min);

d_2 is the outer diameter of shaft (mm);

d_1 is the inner diameter of shaft (mm), taken as 0 for a solid shaft;

L is the spacing between fulcrums of the shaft (m).

4) The maximum deflection of shaft should not exceed $L/3000$. The maximum deflection of a shaft with gears should not exceed 0.03 times the gear module. The maximum deflection angle caused by deflection at the fulcrum should not exceed 0.001 rad. The torsion angle per meter of shaft length should not exceed 0.5°.

8 Structural Design

8.1 Calculation Principles

8.1.1 The allowable stress method shall be used for structural calculation. The strength, stiffness and stability shall be calculated without considering the material plasticity, and the fatigue strength need not be calculated.

8.1.2 The structural calculation shall be conducted according to two types of load combinations. For Type I load combination, the strength, stiffness and stability shall be calculated assuming that the structure bears the maximum working load; for Type II load combination, the strength and stability shall be checked assuming that the structure bears the maximum non-working load or the special working load.

8.1.3 The calculation of stiffness and stability of structural member shall comply with the current sector standard NB/T 10341.1, *Code for Design of Hoists for Hydropower Projects—Part 1: Code for Design of Fixed Wire Rope Hoists*.

8.2 Load Combinations

8.2.1 Load combinations used for calculation shall be in accordance with Table 8.2.1.

Table 8.2.1 Load combinations

Loads	Type I	Type II
Dead load	√	√
Hoist load	√	–
Seismic load	–	√

8.2.2 Installation loads may be added to the existing combinations when necessary.

8.3 Allowable Stresses

8.3.1 The size grouping and yield strength of commonly used structural materials shall be in accordance with Table 8.3.1-1, and the allowable stresses of structural materials commonly used for Type I load combination shall be in accordance with Table 8.3.1-2.

Table 8.3.1-1 Size grouping and yield strength of commonly used structural materials

Group	Q235		Q355	
	Steel thickness (mm)	R_{eH} (N/mm^2)	Steel thickness (mm)	R_{eH} (N/mm^2)
Group 1	≤ 16	235	≤ 16	355
Group 2	> 16 - 40	225	> 16 - 40	345
Group 3	> 40 - 60	215	> 40 - 63	335
Group 4	> 60 - 100	205	> 63 - 80	325

NOTES: R_{eH} shall be looked up in the relevant standard regarding materials when the thickness exceeds the upper limit listed in the table.

Table 8.3.1-2 Allowable stresses of commonly used structural materials for Type Ⅰ load combination (N/mm^2)

Stress type	Symbol	Q235				Q355			
		Group 1	Group 2	Group 3	Group 4	Group 1	Group 2	Group 3	Group 4
Tensile, compressive and bending stress	$[\sigma]$	160	150	145	140	240	235	225	220
Shear stress	$[\tau]$	95	90	85	80	140	135	130	125
Local end face compressive stress	$[\sigma_{cd}]$	225	210	200	195	335	325	315	305
Local contact compressive stress	$[\sigma_{cj}]$	120	115	110	105	180	175	170	165

NOTES:
1 Local compressive stress refers to such states in which a small part of surface of web of the component is being compressed by local load or the end face is bearing pressure.
2 Local contact compressive stress refers to the compressive stress for small-movability hinge on the projection plane of the contact surface.

8.3.2 The allowable stress of welds shall meet the following requirements:

1   In the design of welded connection, the weld shall have the same comprehensive mechanical properties as the base metal.

2   When checking calculation is carried out by Type I load combination, the allowable stresses of welds of commonly used structural materials for Type I load combination shall be in accordance with Table 8.3.2.

Table 8.3.2 Allowable stresses of welds of commonly used structural materials for Type I load combination (N/mm²)

Weld type	Stress type		Symbol	Q235				Q355			
				Group 1	Group 2	Group 3	Group 4	Group 1	Group 2	Group 3	Group 4
Butt weld	Compressive or tensile stress	Class I or II	$[\sigma_h]$	160	150	145	140	240	235	225	220
		Class III	$[\sigma_h]$	130	120	115	110	190	185	180	175
	Shear stress	Class I or II	$[\tau_h]$	95	90	85	80	140	135	130	125
		Class III	$[\tau_h]$	80	75	70	65	110	110	105	100
Fillet weld	Compressive, tensile or shear stress		$[\tau_h]$	115	105	100	100	170	165	160	155

NOTES:
1   The classification of welds shall comply with the current sector standard NB/T 35051, *Code for Manufacture Erection and Acceptance of Gate Hoists in Hydropower Projects*.
2   The allowable stress of overhead weld is multiplied by 0.8 on the basis of the value in Table 8.3.2.
3   The allowable stress of installation weld is multiplied by 0.9 on the basis of the value in Table 8.3.2.
4   When a single-angle steel weldment is connected with one leg, for the equal angle steel, either leg can be used; for the unequal angle steel, the shorter leg can be used, the allowable stress of the connecting weld is multiplied by 0.85.

8.3.3 The allowable stresses of bolt and pin connections shall meet the following requirements:

1   The allowable stresses of bolts shall be selected according to the grade of bolt joint, which is classified as Grade A, Grade B or Grade C.

2   The allowable stresses of bolt and pin connections made of commonly used structural materials for Type I load combination shall be in accordance with Table 8.3.3.

Table 8.3.3 Allowable stresses of bolt and pin connections of commonly used structural materials for Type I load combination (N/mm²)

Connection type	Stress type	Symbol	Performance level of bolt				Steel grade of member							
							Q235				Q355			
			L4.6	L4.8	L5.6	L8.8	Group 1	Group 2	Group 3	Group 4	Group 1	Group 2	Group 3	Group 4
Grade A and Grade B bolt connection (Type I holes)	Tensile stress	$[\sigma]$	–	–	155	315	–	–	–	–	–	–	–	–
	Shear stress	$[\tau]$	–	–	115	235	–	–	–	–	–	–	–	–
	Compressive stress	$[\sigma_c]$	–	–	–	–	290	270	260	250	430	420	405	395
Grade C bolt connection	Tensile stress	$[\sigma]$	125	165	–	–	–	–	–	–	–	–	–	–
	Shear stress	$[\tau]$	95	125	–	–	–	–	–	–	–	–	–	–
	Compressive stress	$[\sigma_c]$	–	–	–	–	225	210	205	200	335	325	315	305
Pin connection (Grade 35 steel)	Bending stress	$[\sigma]$	135				–	–	–	–	–	–	–	–
	Shear stress	$[\tau]$	80				–	–	–	–	–	–	–	–
	Compressive stress	$[\sigma_c]$	–				225	210	205	200	335	325	315	305

NOTES:
1 The following holes shall be taken as Class I holes: the hole drilled in an individual part or member with a drill die to the design aperture; the hole drilled in an assembly to the design aperture; the hole drilled or punched to a small aperture in an individual part and then counter bored in the assembly to the design aperture.
2 The allowable compressive stress of pin should be appropriately reduced when the pin moves slightly during operation.

8.3.4 For Type Ⅱ load combination, the allowable stresses shown in Table 8.3.1-2, Table 8.3.2 and Table 8.3.3 shall be increased by 15 %.

8.3.5 The allowable stresses of grey iron shall be in accordance with Table 8.3.5.

Table 8.3.5　Allowable stresses of grey iron (N/mm²)

Stress Type	Symbol	Grade of grey iron		
		HT150	HT200	HT250
Axial compressive and bending compressive stresses	$[\sigma_a]$	120	150	200
Bending tensile stress	$[\sigma_w]$	35	45	60
Shear stress	$[\tau]$	25	35	45
Local end face compressive stress	$[\sigma_{cd}]$	170	210	260
Local contact compressive stress	$[\sigma_{cj}]$	60	75	90

8.3.6 The allowable compressive stress of phase Ⅰ and Ⅱ concrete for embedded parts shall be in accordance with Table 8.3.6.

Table 8.3.6　Allowable compressive stress of concrete (N/mm²)

Stress Type		Strength class of concrete				
		C15	C20	C25	C30	C40
Compressive stress	$[\sigma_h]$	5	7	9	11	14

8.4 Strength Calculation for Structural Members and Connections

8.4.1 The strength calculation for structural members shall meet the following requirements:

1　When the structural member is in tension, compression, bending or torsion, its strength may be calculated according to the general strength calculation formulae, and the calculated stress shall be less than the allowable stress.

2　When the point load is applied on the upper flange of the beam, the local compressive stress of the web shall be calculated by the following formulae:

$$\sigma_m = \frac{F_p}{\delta(a+2h_y)} \tag{8.4.1-1}$$

$$\sigma_m \leq [\sigma] \tag{8.4.1-2}$$

where

σ_m is the local compressive stress (N/mm^2);

F_p is the point load (N);

δ is the web thickness (mm);

a is the length of point load action (mm);

h_y is the distance from the top surface of the member to the upper edge of the calculation height of the web (mm);

$[\sigma]$ is the allowable stress of steel (N/mm^2).

3   The calculation of combined stress shall meet the following requirements:

1) When the normal stresses σ_x and σ_y and shear stress τ_{xy} in two directions are applied on the same calculation point of a member, the combined stress σ_d at this point shall be calculated by the following formulae:

$$\sigma_d = \sqrt{\sigma_x^2 + \sigma_y^2 - \sigma_x\sigma_y + 3\tau_{xy}^2} \tag{8.4.1-3}$$

$$\sigma_d \leq 1.1[\sigma] \tag{8.4.1-4}$$

where

σ_d is the combined stress at the calculation point (N/mm^2);

σ_x, σ_y are the normal stresses in two directions applied on the same calculation point of a member, with their respective positive or negative sign (N/mm^2), both of which shall be less than the allowable stress $[\sigma]$ in Table 8.3.1-2 of this code;

τ_{xy} is the shear stress on the same calculation point of a member (N/mm^2), which shall be less than $[\tau]$ in Table 8.3.1-2 of this code;

$[\sigma]$ is the allowable stress of steel (N/mm^2).

2) When the normal stress σ, shear stress τ and local compressive stress σ_m are applied on the same calculation point of a member, the combined stress σ_d on this point shall be calculated by the following formulae:

$$\sigma_d = \sqrt{\sigma^2 + \sigma_m^2 - \sigma\sigma_m + 3\tau^2} \tag{8.4.1-5}$$

$$\sigma_d \leq 1.1[\sigma] \qquad (8.4.1\text{-}6)$$

where

- σ_d is the combined stress at the calculation point (N/mm^2);
- σ is the normal stress (N/mm^2) with positive or negative sign;
- σ_m is the local compressive stress (N/mm^2) with positive or negative sign;
- τ is the shear stress (N/mm^2);
- $[\sigma]$ is the allowable stress of steel (N/mm^2).

3) When only the tensile or compressive stress σ and shear stress τ are applied, the combined stress σ_d shall be calculated by the following formulae:

$$\sigma_d = \sqrt{\sigma^2 + 3\tau^2} \qquad (8.4.1\text{-}7)$$

$$\sigma_d \leq 1.1[\sigma] \qquad (8.4.1\text{-}8)$$

where

- σ_d is the combined stress at the calculation point (N/mm^2);
- σ is the tensile or compressive stress (N/mm^2);
- τ is the shear stress (N/mm^2);
- $[\sigma]$ is the allowable stress of steel (N/mm^2).

8.4.2 The strength calculation of welded connection shall meet the following requirements:

1. The stress of butt weld is calculated by the minimum plate thickness t in the connection. When it is not possible to weld with a run-off plate, the calculation length of each weld is taken as its actual length minus $2t$. The calculation of butt welds shall be conducted as follows:

    1) For the butt weld subjected to axial tension or compression, the calculated transverse tensile and compressive stresses shall be less than the allowable values shown in Table 8.3.2 of this code.

    2) For the butt weld subjected to both bending moment and shear stress, the maximum normal stress at the dangerous point shall not be greater than the allowable transverse tensile and compressive stresses of welds shown in Table 8.3.2 of this code, and the maximum shear stress shall not be greater than the allowable shear stress of welds shown in Table 8.3.2 of this code.

3) For the butt weld subjected to combined stress, the combined stress at the position where both normal stress and shear stress are relatively high shall be calculated by the following formulae:

$$\sigma_h = \sqrt{\sigma^2 + 3\tau^2} \qquad (8.4.2\text{-}1)$$

$$\sigma_h \leq 1.1[\sigma_h] \qquad (8.4.2\text{-}2)$$

where

σ_h is the combined stress of butt weld (N/mm^2);

σ is the tensile or compressive stress (N/mm^2);

τ is the shear stress (N/mm^2);

$[\sigma_h]$ is the allowable stress of weld (N/mm^2), which shall be taken in accordance with Table 8.3.2 of this code.

2   The calculation of fillet welds shall meet the following requirements:

1) Under the action of tensile, compressive and shear stresses through the centroid of a weld joint, the stress σ_h of the front fillet weld perpendicular to the weld length direction shall be calculated by the following formulae:

$$\sigma_h = \frac{F}{h_e l_w} \qquad (8.4.2\text{-}3)$$

$$\sigma_h \leq [\tau_h] \qquad (8.4.2\text{-}4)$$

2) Under the action of tensile, compressive and shear forces through the centroid of a weld joint, the stress τ_h of the side fillet weld parallel to the weld length direction shall be calculated by the following formulae:

$$\tau_h = \frac{F}{h_e l_w} \qquad (8.4.2\text{-}5)$$

$$\tau_h \leq [\tau_h] \qquad (8.4.2\text{-}6)$$

3) Under the action of various forces, the stress at the position where both σ_h and τ_h appear shall satisfy the following formula:

$$\sqrt{\sigma_h^2 + \tau_h^2} \leq [\tau_h] \qquad (8.4.2\text{-}7)$$

where

σ_h is the stress perpendicular to the length direction of the weld, which is calculated by the effective section of the weld (N/mm^2);

τ_h is the stress parallel to the length direction of the weld, which is calculated by the effective section of the weld (N/mm²);

F is the force acting on the fillet weld (N);

h_e is the calculation height of the fillet weld (mm), which is taken as $0.7h_f$ for a right-angle fillet weld, and h_f is the smaller leg size, as determined in Figure 8.4.2;

l_w is the calculation length of the fillet weld (mm);

$[\tau_h]$ is the allowable shear stress of the fillet weld, which is taken as per Table 8.3.2 of this code.

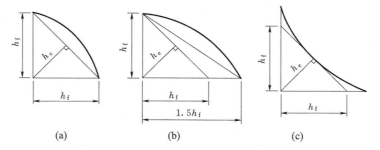

Figure 8.4.2 Sections of right-angle fillet weld

3   The calculation length of a weld shall meet the following requirements:

   1) When the internal force is distributed along the full length of a side fillet weld, the calculation length of the weld is taken as its actual length minus $2h_f$.

   2) The minimum calculation length of a fillet weld is $8h_f$.

   3) For a side fillet weld, the maximum calculation length is $40h_f$ under dynamic load and $60h_f$ under static load. If the weld length is greater than that specified above, the excess need not be considered in the calculation.

8.4.3 The strength calculation of bolt connections shall meet the following requirements:

   1   The general purpose bolt connection shall meet the following requirements:

   1) Grade C bolt connections may only be used for tension connection or temporary fixation during installation as the fit clearance between the bolt and the hole is relatively large.

   2) Grade A or Grade B bolt connections may be used in structures under dynamic load.

        3) For general purpose bolt connections, the tensile capacity and shear capacity of bolts shall be checked, and the compressive capacity of bolt holes shall be checked.

    2   When the hoist frame adopts high-strength bolt connection, the calculation of the bolt connection shall comply with the current national standard GB/T 3811, *Design Rules for Cranes*.

8.4.4   When the foundation embedded parts of the hoist frame are in phase I or II concrete, the compressive stress of the concrete shall be calculated based on the effective compression area of the embedded parts and shall be in accordance with Table 8.3.6 of this code.

9 Electrical Design

9.1 General Requirements

9.1.1 The electrical equipment of electric or manual-electric screw hoists shall comply with the current national standard GB/T 5226.32, *Electrical Safety of Machinery—Electrical Equipment of Machines—Part 32: Requirements for Hoisting Machines.*

9.1.2 Electric wires, cables and their laying shall comply with the current national standard GB/T 3811, *Design Rules for Cranes.*

9.2 Power Supply

9.2.1 The screw hoist should adopt AC power supply with a rated voltage of 380 V.

9.2.2 Under normal working conditions, the voltage fluctuation of power supply system at the feeder access point of the hoist shall not exceed ±10 % of the rated voltage. When the motors start up, the voltage loss from the low-voltage busbar of the supply transformer to any motor terminal of the hoist shall not exceed 15 % of the rated voltage.

9.3 Control and Protection Devices

9.3.1 Electric or manual-electric screw hoists shall be provided with corresponding electrical control system according to the operating conditions and functional requirements.

9.3.2 The local electrical control system should adopt a programmable logic controller (PLC).

9.3.3 When remote control is required, an interface shall be set accordingly in the local electrical control system.

9.3.4 The electrical control system should be provided with electrical protection devices, such as short circuit protection, overcurrent protection, voltage-loss protection, earthing protection, phase fault or open-phase protection, overheating protection, and position limit protection, as well as the main circuit breaker and, if necessary, the emergency switch for disconnecting the master power supply.

9.3.5 For screw hoists used in humid tropics, dry tropics or high altitude areas, the design and type selection of electrical devices shall meet the relevant requirements.

9.4 Lighting and Signaling

9.4.1 The screw hoist room, operating deck, walkway, and stairs shall be provided with appropriate lighting, and the illuminance shall meet the relevant requirements.

9.4.2 The operating deck shall be provided with noticeable master power supply on-off status signaling, equipment fault signaling and alarming devices. The signals may be audible and visual.

9.5　**Earthing and Lightning Protection**

9.5.1 Earthing protection shall meet the following requirements:

1　The metal structure of the hoist shall be reliably connected to the earthing network.

2　The metal enclosures, conduits, brackets and trunking of all the electrical equipment of a hoist shall be reliably earthed. Dedicated earthing conductors should be used to ensure reliable earthing of the electrical equipment.

3　The conductance of earthing conductors and earthing facilities should not be less than 1/2 of the maximum phase conductance in the line.

4　The earthing conductors shall not be used as current-carrying neutral lines.

9.5.2 For hoists installed outdoors and higher than the surrounding ground surface, consideration shall be given to avoiding damage to high-position parts and personal injury caused by lightning strike.

10 Safety

10.1 Markings, Nameplates and Safety Signs

10.1.1 Hoists shall be provided with markings, nameplates and safety signs.

10.1.2 The rated hoisting capacity shall be permanently marked at a conspicuous position.

10.1.3 Each hoist shall be provided with a nameplate at an appropriate position, indicating the following as a minimum:

1. Name of manufacturer.
2. Product name and model.
3. Main performance parameters.
4. Ex-factory serial number.
5. Date of manufacture.

10.1.4 Legible textual safety signs shall be provided at appropriate positions of the hoist. Safety signs shall be provided at the hazardous positions of the hoist. The safety signs shall comply with the current national standards GB 2894, *Safety Signs and Guideline for the Use*; and GB/T 15052, *Cranes—Safety Signs and Hazard Pictorials—General Principles*. The colours of safety signs shall comply with the current national standard GB 2893, *Safety Colours*.

10.2 Safety Requirements for Mechanism, Structure, Electrical and Fire Protection

10.2.1 The operating deck shall meet the following requirements:

1. The operating deck shall have sufficient space to meet the ergonomic requirements.
2. The operating deck shall be provided with an access to operate, inspect and maintain the hoist, and the ladder and deck shall be provided with guardrails.
3. The floor of the operating deck shall be covered by non-slip adiabatic non-metallic material.
4. The illuminance on the working face of the operating deck shall not be less than 30 Lx.
5. Important operation indicators shall provide conspicuous displays, and be installed at the positions convenient for operators to observe. The

indicators, alarm lights and emergency stop switch button shall have durable and legible markings. The indicators shall have appropriate ranges and be easy to read. The alarm lights shall have appropriate colors, and red lights shall be used for hazard indications.

10.2.2 Protection shall meet the following requirements:

1 Under normal operation or maintenance conditions of the hoist, protective devices shall be set to prevent foreign matters from getting in, or prevent operators from injury by the running parts. Protective covers or guardrails shall be provided for exposed moving parts of the hoist such as couplings and transmission shafts, which might injure operators.

2 Rainproof measures shall be taken for the electrical equipment of outdoor hoists.

10.2.3 The control and operation system shall meet the following requirements:

1 The design and layout of the control and operation system shall be able to avoid mis-operation, and to ensure the safe and reliable operation of the hoist in normal service.

2 Text marks or symbols shall be attached on or near each control device to distinguish the functions and clearly indicate the movement direction of the hoist parts.

10.2.4 Fire protection facilities shall be provided according to the project rank. The configuration of fire protection facilities shall comply with the current national standard GB 50872, *Code for Fire Protection Design of Hydropower Projects*.

10.3 Operation and Maintenance Manuals

10.3.1 The hoist supplier shall provide the user with the Hoist Driving Manual to guide the safe use of the hoist, and its content shall comply with the current national standard GB/T 17909.1, *Cranes—Crane Driving Manual—Part 1: General*.

10.3.2 The hoist supplier shall provide the user with the Hoist Maintenance Manual to guide the maintenance of the hoist, and its content shall comply with the current national standard GB/T 18453, *Cranes—Maintenance Manual—Part 1: General*.

Appendix A Models and Series Parameters of Screw Hoists

A.0.1 The designation of screw hoist model should be as shown in Figure A.0.1:

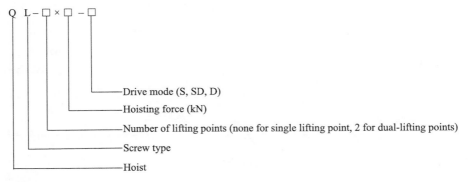

Key

S   manual
SD   manual-electric
D   electric

Figure A.0.1 Designation of screw hoist model

A.0.2 The parameters of screw hoist series should be in accordance with Table A.0.2.

Table A.0.2 Parameters of screw hoist series

Type	Model	Lifting force (kN)	Closing force (kN)	Stroke (m)	Hoisting speed (electric) (m/min)	Center distance between lifting points (m)
Single lifting point hoist	QL-10-S	10	5	1.5	—	—
	QL-16-S	16	8	2.0		
	QL-20-S	20	10	2.0		
	QL-25-S	25	12.5	2.5		
	QL-32-S	32	16	2.5		
	QL-40-S	40	20	3.0		
	QL-50-S	50	25	3.0		
	QL-63-S	63	32	3.5		

Table A.0.2 *(continued)*

Type	Model	Lifting force (kN)	Closing force (kN)	Stroke (m)	Hoisting speed (electric) (m/min)	Center distance between lifting points (m)
Single lifting point hoist	QL-80-S	80	40	3.5	–	–
	QL-100-S	100	50	4.0		
	QL-125-S	125	63	4.0		
	QL-160-S	160	80	4.5		
	QL-200-S	200	100	4.5		
	QL-10-SD	10	5	1.5	0.1 - 0.5	–
	QL-16-SD	16	8	2.0		
	QL-20-SD	20	10	2.0		
	QL-25-SD	25	12.5	2.5		
	QL-32-SD	32	16	2.5		
	QL-40-SD	40	20	3.0		
	QL-50-SD	50	25	3.0		
	QL-63-SD	63	32	3.5		
	QL-80-SD	80	40	3.5		
	QL-100-SD	100	50	4.0		
	QL-125-SD	125	63	4.0		
	QL-160-SD	160	80	4.5		
	QL-200-D	200	100	4.5		
	QL-250-D	250	125	4.5		
	QL-320-D	320	160	4.5		
	QL-400-D	400	200	5.0		
	QL-500-D	500	250	5.5		
	QL-630-D	630	320	5.5		
	QL-800-D	800	400	6.0		
	QL-1000-D	1000	500	8.0		

Table A.0.2 *(continued)*

Type	Model	Lifting force (kN)	Closing force (kN)	Stroke (m)	Hoisting speed (electric) (m/min)	Center distance between lifting points (m)
Dual-lifting point hoist	QL-2×25-SD	2 × 25	2 × 12.5	2.5	0.1 - 0.5	1.0 - 3.0
	QL-2×40-SD	2 × 40	2 × 20	3.0		1.0 - 3.0
	QL-2×50-SD	2 × 50	2 × 25	3.0		1.0 - 3.0
	QL-2×63-SD	2 × 63	2 × 32	3.5		1.0 - 3.0
	QL-2×80-SD	2 × 80	2 × 40	3.5		1.0 - 3.0
	QL-2×100-SD	2 × 100	2 × 50	4.0		1.0 - 3.0
	QL-2×125-SD	2 × 125	2 × 63	4.0		1.0 - 3.0
	QL-2×160-SD	2 × 160	2 × 80	4.5		1.5 - 3.5
	QL-2×200-D	2 × 200	2 × 100	4.5		1.5 - 3.5
	QL-2×250-D	2 × 250	2 × 125	4.5		1.5 - 3.5
	QL-2×320-D	2 × 320	2×160	4.5		1.5 - 3.5
	QL-2×400-D	2 × 400	2×200	5.0		2.0 - 4.0
	QL-2×500-D	2 × 500	2×250	5.5		2.0 - 4.0
	QL-2×630-D	2 × 630	2×320	5.5		2.0 - 4.0
	QL-2×800-D	2 × 800	2×400	6.0		2.0 - 4.0
	QL-2×1000-D	2 × 1000	2×500	8.0		3.0 - 5.0

NOTES:
1. The upper limit of center distance between lifting points refers to the allowable distance between lifting points when there is no support for the synchronizing shaft of the hoist.
2. When the stroke exceeds the value listed in the table, the closing force shall be lowered as appropriate.

A.0.3 Screw hoists should be classified, by the hoisting force acting on a single lifting point, into: small-sized (< 250 kN), medium-sized (≥ 250 kN but < 500 kN), and large-sized (≥ 500 kN).

Appendix B Design Profiles and Dimensions of Trapezoidal Screw Threads

B.0.1 The design profile, items, symbols and relational expressions of a trapezoidal screw thread shall be in accordance with Table B.0.1.

Table B.0.1 Design profile, items, symbols and relational expressions of a trapezoidal screw thread

Schematic diagram of profile	Item	Symbol	Relational expression
(see figure below)	Clearance at major and minor diameters	a_c	
	Major diameter of internal thread of design profile	D_4	$D_4 = d + 2a_c$
	Pitch diameter of internal thread of design profile	D_2	
	Minor diameter of internal thread of design profile	D_1	$D_1 = d - 2H_1 = d - P$
	Major diameter of external thread of design profile (nominal diameter)	d	
	Pitch diameter of external thread of design profile	d_2	$d_2 = D_2 = d - H_1 = d - 0.5P$
	Minor diameter of external thread of design profile	d_3	$d_3 = d - 2h_3 = d - P - a_c$
	Thread height	H_1	$H_1 = 0.5P$
	Internal thread height of design profile	H_4	$H_4 = h_3 = H_1 + a_c = 0.5P + a_c$
	External thread height of design profile	h_3	$h_3 = H_1 + a_c = 0.5P + a_c$
	Thread pitch	P	
	Maximum radius on crest corners of external thread on design profile	$R_{1\max}$	$R_{1\max} = 0.5a_c$
	Maximum radius on root corners of internal and external threads on design profile	$R_{2\max}$	$R_{2\max} = a_c$

1—Internal thread; 2—External thread

B.0.2 The dimensions of design profiles of trapezoidal screw threads shall be in accordance with Table B.0.2.

Table B.0.2　Dimensions of design profiles of trapezoidal screw threads (mm)

Pitch P	a_c	$H_4 = h_3$	R_{1max}	R_{2max}
1.5	0.15	0.90	0.075	0.15
2	0.25	1.25	0.125	0.25
3	0.25	1.75	0.125	0.25
4	0.25	2.25	0.125	0.25
5	0.25	2.75	0.125	0.25
6	0.5	3.5	0.25	0.5
7	0.5	4	0.25	0.5
8	0.5	4.5	0.25	0.5
9	0.5	5	0.25	0.5
10	0.5	5.5	0.25	0.5
12	0.5	6.5	0.25	0.5
14	1	8	0.5	1
16	1	9	0.5	1
18	1	10	0.5	1
20	1	11	0.5	1
22	1	12	0.5	1
24	1	13	0.5	1
28	1	15	0.5	1
32	1	17	0.5	1
36	1	19	0.5	1
40	1	21	0.5	1
44	1	23	0.5	1

Appendix C Friction Factors and Efficiencies

C.0.1 The friction factor should be in accordance with Table C.0.1.

Table C.0.1 Friction factor

Friction pair type		Friction factor
Sliding bearing		0.1
Rolling bearing	Ball bearing or roller bearing	0.015
	Taper roller bearing	0.02

C.0.2 The efficiency should be in accordance with Table C.0.2.

Table C.0.2 Efficiency

Transmission part		Efficiency	
		Sliding bearing	Rolling bearing
Cylindrical gear drive	Open cylindrical gear pair (grease lubrication)	0.90 - 0.92	0.92 - 0.94
	Enclosed cylindrical gear pair (oil lubrication)		0.96 - 0.98
Bevel gear drive	Open bevel gear pair (grease lubrication)	0.90 - 0.92	0.92 - 0.94
	Enclosed bevel gear pair (oil lubrication)		0.95 - 0.97
Intermediate shaft		0.95 - 0.97	0.97 - 0.99
Gear coupling		0.99	

Appendix D Stability Coefficient of Eccentric Compression

D.0.1 The stability coefficient of eccentric compression shall be in accordance with Table D.0.1.

Table D.0.1 Stability coefficient of eccentric compression

λ	ε						
	0	0.2	0.4	0.6	0.8	1.0	1.2	1.4
0	1.000	0.930	0.875	0.819	0.766	0.720	0.675	0.630
10	0.995	0.920	0.855	0.795	0.742	0.695	0.648	0.610
20	0.981	0.900	0.826	0.766	0.710	0.662	0.620	0.533
30	0.958	0.875	0.795	0.730	0.680	0.630	0.591	0.555
40	0.927	0.830	0.753	0.688	0.635	0.597	0.560	0.526
50	0.888	0.788	0.712	0.647	0.598	0.558	0.524	0.492
60	0.842	0.736	0.668	0.606	0.560	0.523	0.491	0.459
70	0.789	0.676	0.618	0.559	0.518	0.482	0.453	0.428
80	0.731	0.630	0.572	0.521	0.480	0.446	0.417	0.393
90	0.669	0.571	0.521	0.477	0.440	0.411	0.388	0.364
100	0.604	0.530	0.478	0.441	0.408	0.379	0.357	0.336
λ	ε							
	1.6	1.8	2.0	2.5	3.0	3.5	4.0	4.5
0	0.596	0.562	0.534	0.468	0.414	0.370	0.333	0.303
10	0.575	0.546	0.518	0.455	0.404	0.362	0.325	0.298
20	0.550	0.520	0.495	0.439	0.390	0.349	0.315	0.288
30	0.525	0.496	0.473	0.420	0.373	0.335	0.303	0.277
40	0.494	0.469	0.449	0.390	0.355	0.320	0.290	0.265
50	0.462	0.436	0.420	0.377	0.338	0.304	0.277	0.253
60	0.433	0.412	0.395	0.355	0.319	0.289	0.263	0.241
70	0.403	0.381	0.370	0.334	0.301	0.273	0.249	0.230
80	0.370	0.358	0.344	0.314	0.283	0.258	0.236	0.218
90	0.347	0.333	0.322	0.294	0.266	0.243	0.224	0.207
100	0.317	0.303	0.292	0.275	0.250	0.229	0.211	0.197

Table D.0.1 *(continued)*

λ	ε								
	5.0	**5.5**	**6.0**	**6.5**	**7.0**	**8.0**	**9.0**	**10**
0	0.277	0.256	0.235	0.220	0.205	0.182	0.162	0.147
10	0.271	0.251	0.231	0.217	0.210	0.179	0.160	0.145
20	0.263	0.243	0.225	0.210	0.196	0.174	0.157	0.141
30	0.254	0.234	0.218	0.203	0.191	0.169	0.152	0.138
40	0.243	0.226	0.210	0.196	0.184	0.164	0.148	0.135
50	0.234	0.216	0.201	0.189	0.177	0.159	0.143	0.130
60	0.224	0.207	0.193	0.182	0.171	0.153	0.138	0.126
70	0.213	0.198	0.185	0.174	0.164	0.147	0.134	0.122
80	0.203	0.189	0.177	0.167	0.157	0.142	0.129	0.118
90	0.192	0.180	0.169	0.160	0.151	0.136	0.124	0.114
100	0.183	0.172	0.161	0.153	0.144	0.131	0.120	0.110

λ	ε					
	12	**14**	**16**	**18**	**20**	**25**	**30**
0	0.123	0.106	0.094	0.084	0.075	0.060	0.050
10	0.122	0.105	0.093	0.083	0.074	0.060	0.050
20	0.120	0.102	0.090	0.080	0.072	0.059	0.049
30	0.117	0.100	0.087	0.078	0.071	0.058	0.048
40	0.114	0.098	0.086	0.077	0.070	0.057	0.047
50	0.111	0.096	0.085	0.075	0.069	0.056	0.046
60	0.107	0.094	0.084	0.074	0.068	0.055	0.045
70	0.104	0.091	0.082	0.073	0.066	0.054	0.044
80	0.101	0.089	0.080	0.072	0.065	0.053	0.043
90	0.098	0.087	0.078	0.070	0.063	0.052	0.042
100	0.095	0.084	0.075	0.068	0.063	0.051	0.042

D.0.2 The eccentricity shall be calculated by the following formula:

$$\varepsilon = \frac{M}{F_2}\frac{A}{W} \tag{D.0.2}$$

where

- ε is the eccentricity;
- M is the additional bending moment (N·mm);
- A is the minor sectional area of the screw rod (mm^2);
- F_2 is the closing force (N);
- W is the resistance moment of minor section of the screw rod (mm^3).

D.0.3 The slenderness ratio shall be calculated by the following formula:

$$\lambda = \frac{4\mu L}{d_1} \tag{D.0.3}$$

where

- λ is the slenderness ratio;
- μ is the conversion coefficient for length;
- L is the actual compression length of the screw rod (mm);
- d_1 is the minor diameter of thread (mm).

Appendix E Motor Power Correction Considering Working Environment

E.0.1 When the motor works at an altitude exceeding 1000 m or an ambient temperature inconsistent with the rated ambient temperature, the motor power shall be corrected by the following formula:

$$N'_N = \frac{N_N}{K} \quad (E.0.1)$$

where

N'_N is the corrected power considering the ambient temperature and altitude, which is used to select the motor (kW);

N_N is the required motor power before correction (kW);

K is the power correction factor, which is determined by Figure E.0.1.

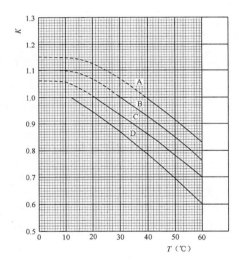

Key

K power correction factor
T ambient temperature
A altitude below 1000 m
B altitude below 2000 m
C altitude below 3000 m
D altitude below 4000 m

Figure E.0.1 Power correction factor as a function of ambient temperature and altitude

E.0.2 When K is greater than 1, its value shall be jointly determined by the motor manufacturer and the hoist manufacturer.

E.0.3 When the altitude exceeds 1000 m, the ambient temperature shall be indicated.

Appendix F Allowable Values of Common Sliding Bearing Materials

F.0.1 Allowable values of copper alloy bearing materials shall be in accordance with Table F.0.1.

Table F.0.1 Allowable values of copper alloy bearing materials

Material of faying surface		$[p]$ (N/mm^2)	$[v]$ (m/s)	$[pv]$ [N·m/(mm^2·s)]
Tin bronze	ZCuSn10P1	15	10	15
	ZCuSn5Pb5Zn5	8	6	6
Cast aluminum bronze	ZCuAl10Fe3	30	8	12
	ZCuAl10Fe3Mn2	20	5	15
Cast lead bronze	ZCuPb30	15	8	60

F.0.2 Self-lubricating bearing materials

For self-lubricating bearing materials such as steel-based copper-plastic composite materials, copper alloy inlaid self-lubricating materials, and engineering plastic alloy materials, the manufacturers shall entrust testing agencies with professional qualification to determine $[p]$, $[v]$ and $[pv]$ by the test methods stipulated in relevant technical codes of China.

Explanation of Wording in This Code

1. Words used for different degrees of strictness are explained as follows in order to mark the differences in executing the requirements in this code:

    1) Words denoting a very strict or mandatory requirement:

        "Must" is used for affirmation; "must not" for negation.

    2) Words denoting a strict requirement under normal conditions:

        "Shall" is used for affirmation; "shall not" for negation.

    3) Words denoting a permission slight choice or an indication of the most suitable choice when conditions permit:

        "Should" is used for affirmation; "should not" for negation.

    4) "May" is used to express the option available, sometimes with the conditional permit.

2. "Shall meet the requirements of …" or "shall comply with …" is used in this code to indicate that it is necessary to comply with the requirements stipulated in other relative standards and codes.

List of Quoted Standards

GB 50009,	*Load Code for the Design of Building Structures*
GB 50872,	*Code for Fire Protection Design of Hydropower Projects*
GB/T 699,	*Quality Carbon Structure Steels*
GB/T 700,	*Carbon Structural Steels*
GB/T 1176,	*Casting Copper and Copper Alloys*
GB/T 1220,	*Stainless Steel Bars*
GB/T 1299,	*Tool and Mould Steels*
GB/T 1231,	*Specifications of High Strength Bolts with Large Hexagon Head, Large Hexagon Nuts, Plain Washers for Steel Structures*
GB/T 1348,	*Spheroidal Graphite Iron Castings*
GB/T 1591,	*High Strength Low Alloy Structural Steels*
GB 2893,	*Safety Colours*
GB 2894,	*Safety Signs and Guideline for the Use*
GB/T 3077,	*Alloy Structure Steels*
GB/T 3098.1,	*Mechanical Properties of Fasteners—Bolts, Screws and Studs*
GB/T 3098.2	*Mechanical Properties of Fasteners—Nuts*
GB/T 3098.3,	*Mechanical Properties of Fasteners—Set Screws*
GB/T 3098.6,	*Mechanical Properties of Fasteners—Stainless Steel Bolts, Screws and Studs*
GB/T 3098.15,	*Mechanical Properties of Fasteners—Stainless Steel Nuts*
GB/T 3811,	*Design Rules for Cranes*
GB/T 4323,	*Pin Coupling with Elastic Sleeve*
GB/T 4942.1,	*Degrees of Protection Provided by the Integral Design of Rotating Electrical Machined (IP Code)—Classification*
GB/T 5117,	*Covered Electrodes for Manual Metal Arc Welding of Non-alloy and Fine Grain Steels*
GB/T 5118,	*Covered Electrodes for Manual Metal Arc Welding of*

	Creep-Resisting Steels
GB/T 5226.32,	*Electrical Safety of Machinery—Electrical Equipment of Machines—Part 32: Requirements for Hoisting Machines*
GB/T 5293,	*Solid Wire Electrodes, Tubular Cored Electrodes and Electrode/Flux Combinations for Submerged Arc Welding of Non Alloy and Fine Grain Steels*
GB/T 5796.3,	*Trapezoidal Screw Threads—Part 3: Basic Dimensions*
GB/T 5796.4,	*Trapezoidal Screw Threads—Part 4: Tolerances*
GB/T 9439,	*Grey Iron Castings*
GB/T 9440,	*Malleable Iron Castings*
GB/T 11352,	*Carbon Steel Castings for General Engineering Purpose*
GB/T 12470,	*Solid Wire Electrodes, Tubular Cored Electrodes and Electrode/Flux Combinations for Submerged Arc Welding of Creep-Resisting Steels*
GB/T 15052,	*Cranes—Safety Signs and Hazard Pictorials—General Principles*
GB/T 17909.1,	*Cranes—Crane Driving Manual—Part 1: General*
GB/T 18254,	*High-Carbon Chromium Bearing Steel*
GB/T 18453,	*Cranes—Maintenance Manual—Part 1: General*
GB/T 20878,	*Stainless and Heat-Resisting Steels—Designation and Chemical Composition*
DL/T 5358,	*Technical Code for Anticorrosion of Metal Structures in Hydroelectric and Hydraulic Engineering*
JB/T 6402,	*Heavy Low Alloy Steel Castings—Technical Specification*
JB/T 6398,	*Heavy Stainless Acid Resistant Steel and Heat Resistant Steel Forgings—Technical Specification*
JB/T 8854.2,	*Curved Tooth Coupling GⅡ CL, GⅡ CLZ*
NB/T 10341.1,	*Code for Design of Hoists for Hydropower Projects—Part 1: Code for Design of Fixed Wire Rope Hoists*
NB/T 35051,	*Code for Manufacture Erection and Acceptance of Gate Hoists in Hydropower Projects*